AF575531

ANIMALS AT RISK

Polar Bears

by Rachel Grack

BLASTOFF! READERS 2

BELLWETHER MEDIA • MINNEAPOLIS, MN

Blastoff! Readers are carefully developed by literacy experts to build reading stamina and move students toward fluency by combining standards-based content with developmentally appropriate text.

Level 1 provides the most support through repetition of high-frequency words, light text, predictable sentence patterns, and strong visual support.

Level 2 offers early readers a bit more challenge through varied sentences, increased text load, and text-supportive special features.

Level 3 advances early-fluent readers toward fluency through increased text load, less reliance on photos, advancing concepts, longer sentences, and more complex special features.

★ **Blastoff! Universe**

Reading Level

Grade K

Grades 1–3

Grade 4

This edition first published in 2022 by Bellwether Media, Inc.

No part of this publication may be reproduced in whole or in part without written permission of the publisher. For information regarding permission, write to Bellwether Media, Inc., Attention: Permissions Department, 6012 Blue Circle Drive, Minnetonka, MN 55343.

Library of Congress Cataloging-in-Publication Data

Names: Koestler-Grack, Rachel A., 1973- author.
Title: Polar bears / Rachel Grack.
Description: Minneapolis, MN : Bellwether Media, 2022. | Series: Animals at risk | Includes bibliographical references and index. | Audience: Ages 5-8 | Audience: Grades 2-3 | Summary: "Simple text and full-color photography introduce beginning readers to the threats to and protections for polar bears. Developed by literacy experts for students in kindergarten through third grade"- Provided by publisher.
Identifiers: LCCN 2021046034 (print) | LCCN 2021046035 (ebook) | ISBN 9781644875902 (library binding) | ISBN 9781648346019 (ebook)
Subjects: LCSH: Polar bear--Juvenile literature. | Polar bear--Behavior--Juvenile literature. | Polar bear--Conservation--Juvenile literature.
Classification: LCC QL737.C27 K63 2022 (print) | LCC QL737.C27 (ebook) | DDC 599.786--dc23
LC record available at https://lccn.loc.gov/2021046034
LC ebook record available at https://lccn.loc.gov/2021046035

Text copyright © 2022 by Bellwether Media, Inc. BLASTOFF! READERS and associated logos are trademarks and/or registered trademarks of Bellwether Media, Inc.

Editor: Kieran Downs Designer: Brittany McIntosh

Printed in the United States of America, North Mankato, MN.

Table of Contents

Cool Bears

Polar bears are large bears that look white. They live in the icy **Arctic**.

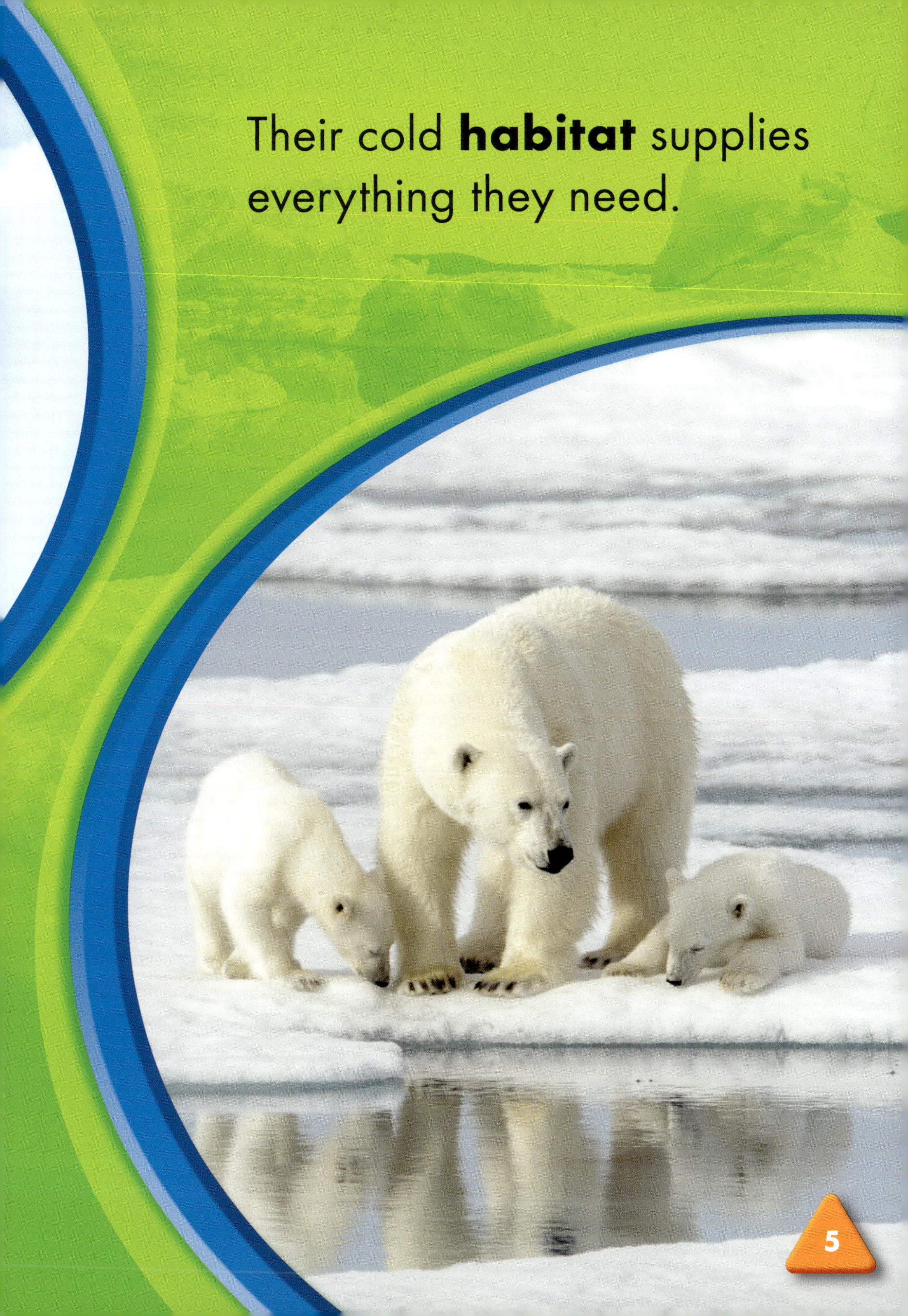

Their cold **habitat** supplies everything they need.

Polar bears spend much of their time on the sea ice.

They search for places to hunt. But sea ice is melting.

In Danger!

Climate change is warming the Arctic quickly. Sea ice is becoming open water.

Without sea ice, polar bears cannot get to food. They will **starve**.

Threats

1 Arctic gets warmer

2 sea ice melts

3 polar bears cannot get to food

Pollution causes other problems for polar bears. It makes their food and water dirty.

Many polar bears are getting sick.

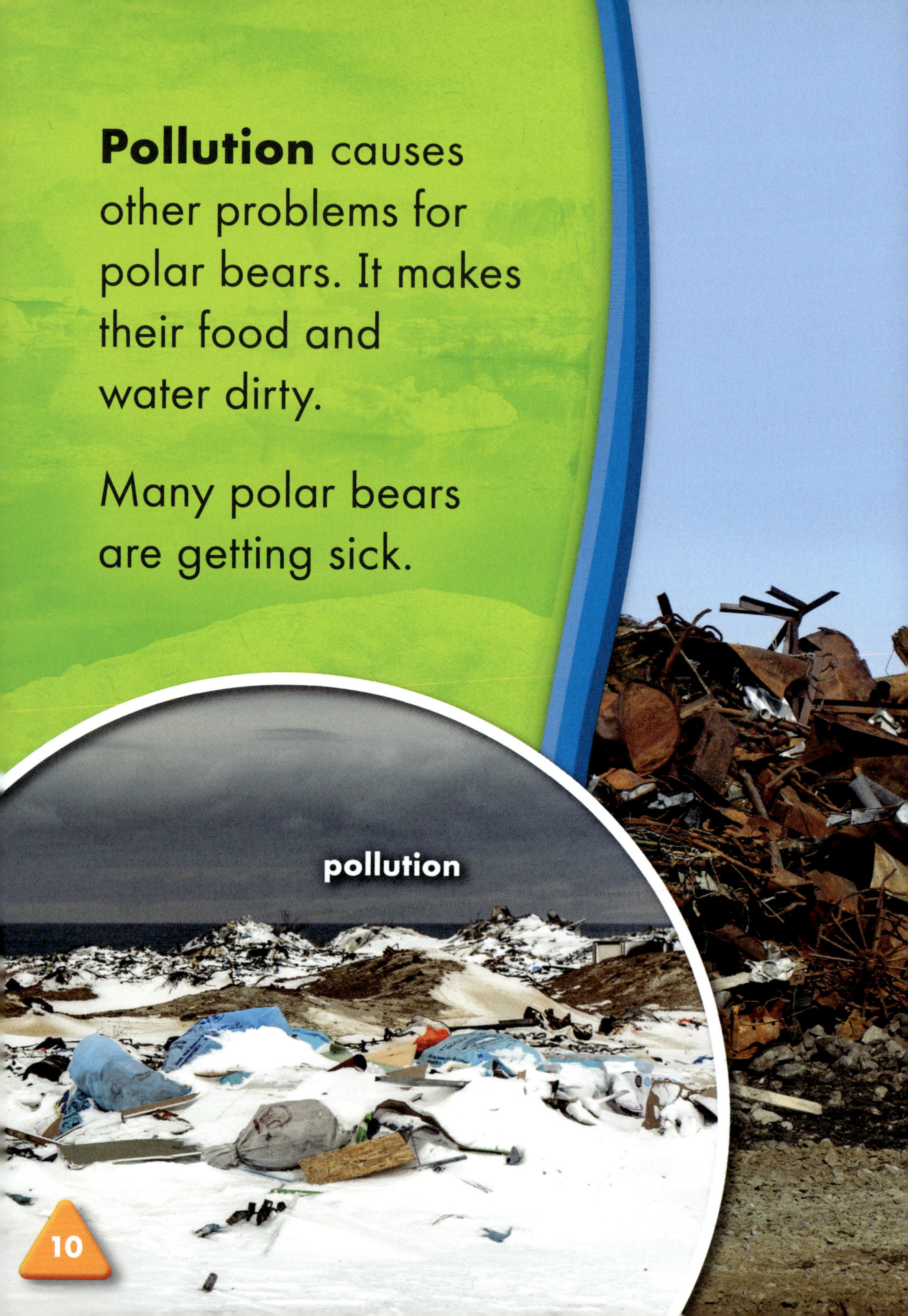

pollution

Polar Bear Stats

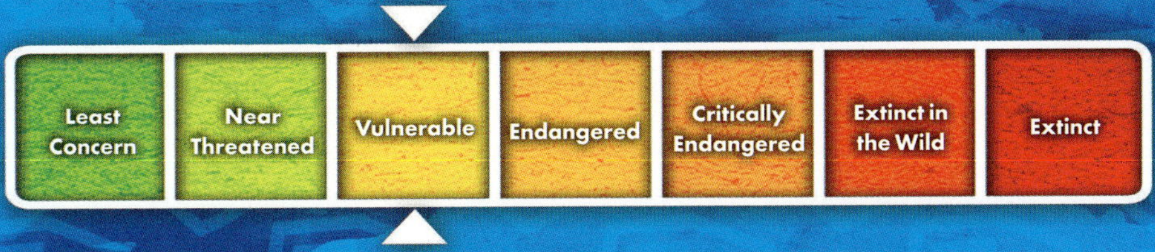

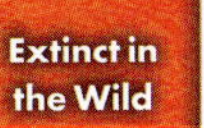

conservation status: vulnerable

life span: about 25 years

Save the Polar Bears!

Polar bears are **apex predators**. They hunt seals and fish. This keeps the Arctic **ecosystem** healthy.

But polar bears could go **extinct** in the wild without help.

The World with Polar Bears

more sea ice

more polar bears

healthy number of seals and fish

Governments are working together to fight climate change. Many countries have agreed to cut **emissions** by certain dates.

This will help slow the loss of sea ice.

Some vehicles and forms of **energy** add to emissions.

Countries are moving toward clean energy, like **solar** and wind power.

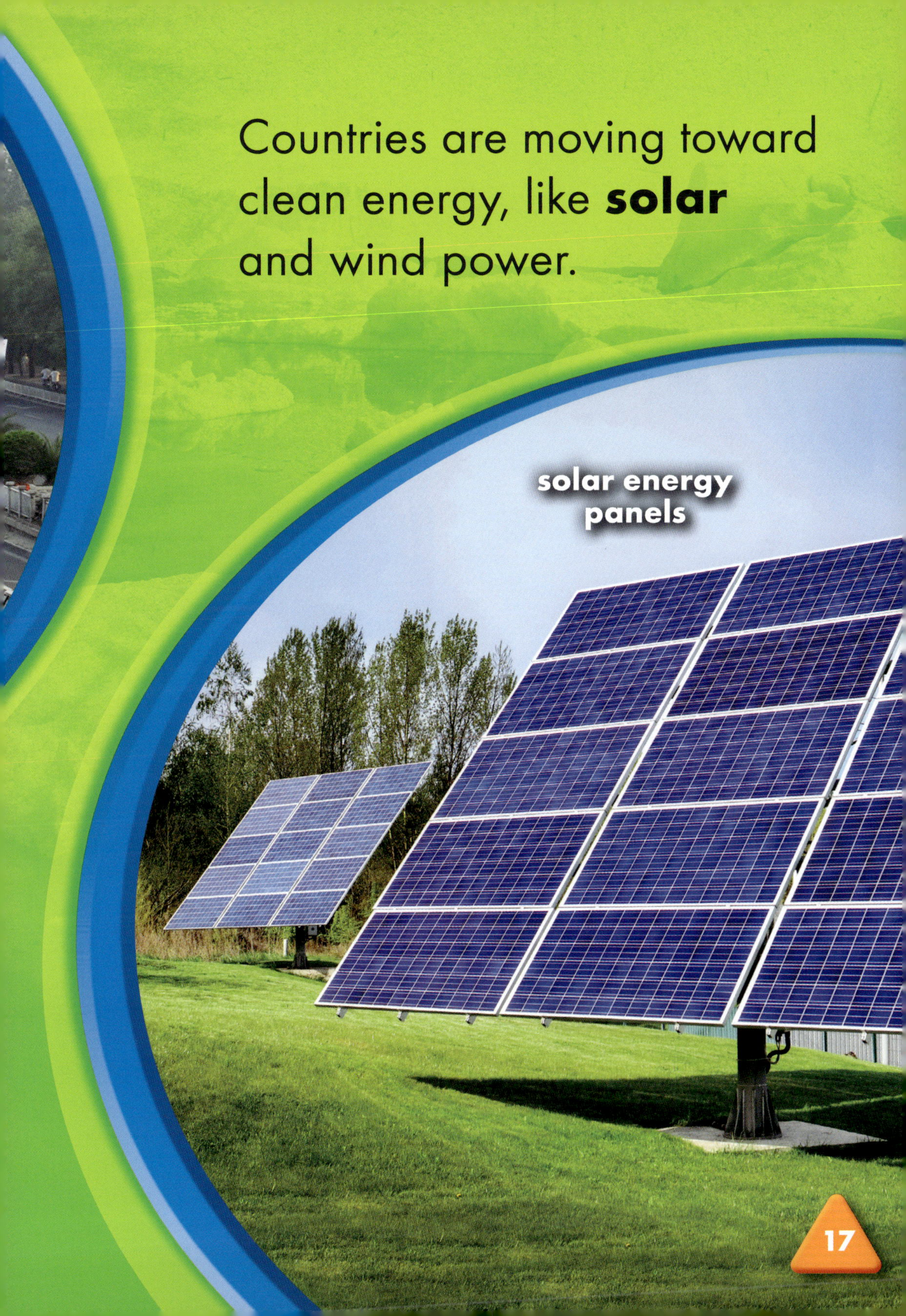

Laws help keep the Arctic safe from businesses that pollute. Oil companies cannot drill in places near polar bears.

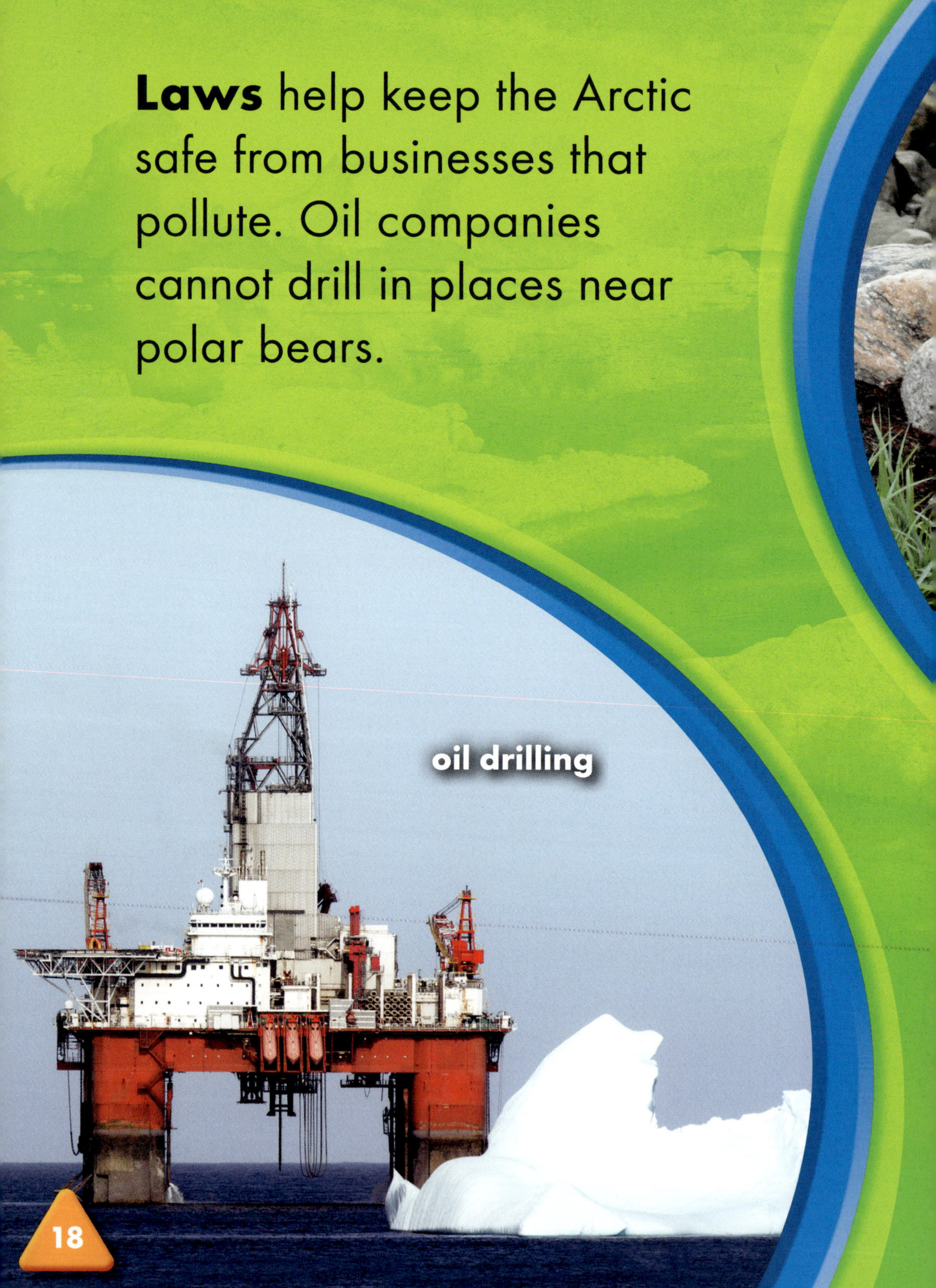

oil drilling

Groups are working to keep ships away from polar bear habitats.

People can help polar bears. They can walk or ride bikes instead of driving.

People can save energy by turning off lights and electronics. Everyone can work to save these cool bears!

Glossary

apex predators—animals at the top of the food chain that are not preyed upon by other animals

Arctic—the cold, frozen land and seas around the North Pole

climate change—a human-caused change in Earth's weather due to warming temperatures

ecosystem—a community of plants and animals that live in a certain place

emissions—substances put into the air that add to climate change

energy—power to make things work

extinct—no longer living

habitat—the place where an animal lives

laws—rules that must be followed

pollution—substances that make nature dirty; pollution usually comes from humans.

solar—related to the sun

starve—to die from lack of food

To Learn More

AT THE LIBRARY

Huddleston, Emma. *Saving Polar Bears.* Lake Elmo, Minn.: Focus Readers, 2021.

Pettiford, Rebecca. *Polar Bears.* Minneapolis, Minn.: Bellwether Media, 2019.

Schuh, Mari. *Polar Bears.* Minneapolis, Minn.: Jump!, 2022.

ON THE WEB

FACTSURFER

Factsurfer.com gives you a safe, fun way to find more information.

1. Go to www.factsurfer.com.
2. Enter "polar bears" into the search box and click 🔍.
3. Select your book cover to see a list of related content.

Index

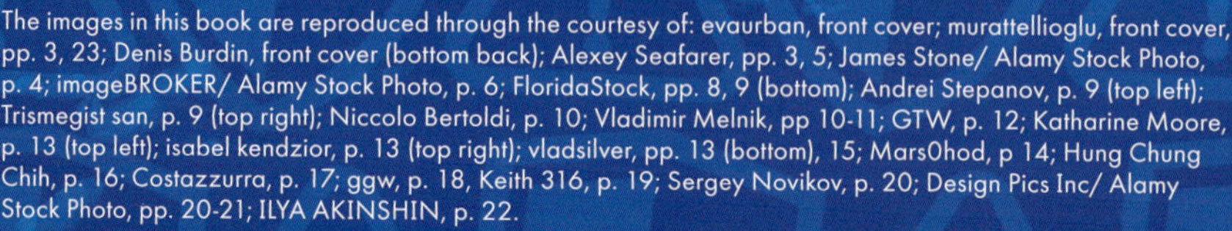

The images in this book are reproduced through the courtesy of: evaurban, front cover; murattellioglu, front cover, pp. 3, 23; Denis Burdin, front cover (bottom back); Alexey Seafarer, pp. 3, 5; James Stone/ Alamy Stock Photo, p. 4; imageBROKER/ Alamy Stock Photo, p. 6; FloridaStock, pp. 8, 9 (bottom); Andrei Stepanov, p. 9 (top left); Trismegist san, p. 9 (top right); Niccolo Bertoldi, p. 10; Vladimir Melnik, pp 10-11; GTW, p. 12; Katharine Moore, p. 13 (top left); isabel kendzior, p. 13 (top right); vladsilver, pp. 13 (bottom), 15; Mars0hod, p 14; Hung Chung Chih, p. 16; Costazzurra, p. 17; ggw, p. 18, Keith 316, p. 19; Sergey Novikov, p. 20; Design Pics Inc/ Alamy Stock Photo, pp. 20-21; ILYA AKINSHIN, p. 22.